Aliens and Secret Technology

## Aliens & Secret Technology-A Theory of the Hidden Truth

This book is an overview of both the history of aliens and UFOS, as well as a review of hidden aerospace and propulsive technologies.

Where did these technologies come from? And how have they developed in the intervening years?

What are some of the Alien races which exist? Do we have a secret space force and are their secret bases on the moon?

Also, I try to answer some questions of how likely some of these claims are—because some things having to do with UFOs and aliens are much more documented than others.

There are also details of many alien related and secret programs that most readers may have never heard of. The most interesting ones are summarized here.

## Copyright Page

The book is copyrighted for 2017

Aliens and Secret Technology—A Theory of the Hidden Truth

By Martin K. Ettington

All Rights Reserved USA 2017

ISBN: 9781973394877

Printed in the United States of America

## Other books by Martin K. Ettington

Spiritual and Metaphysics Books:
Prophecy: A History and How to Guide
God Like Powers and Abilities
Enlightenment for Newbies
Removing Illusions to Find True Happiness
Using the Scientific Method to Study the Paranormal
A Compendium of Metaphysics and How to Guides (Six books together in one volume)
Love from the Heart
The Enlightenment Experience
Learn Your Soul's Purpose
Pursuing Enlightenment
A Modern Man's Search for Truth
Use Intuition and Prophecy to Improve Your Life
The Handbook of Spiritual and Energy Healing

Longevity & Immortality:
Physical Immortality: A History and How to Guide
The Commentaries of Living Immortals
Records of Extremely Long Lived Persons
Enlightenment and Immortality
Longevity Improvements from Science
The 10 Principles of Personal Longevity
Telomeres & Longevity
The Diets and Lifestyles of the Worlds Oldest Peoples
The Longevity Six Books Bundle

Science Fiction:
Out of This Universe
Personal Freedom-Parts 1 & 2
The Psychic Soldier Series:
   Book 1-Himalayan Journey
   Book 2-A Soldier is Born
   Book 3-Fighting For Right
   Book 4-Earth Protector
The Immortality Sci Fi Bundle

The God Like Powers Series:
Human Invisibility
Invulnerability and Shielding
Teleportation
Psychokinesis
Our Energy Body, Auras, and Thoughtforms

The God Like Powers Series—
   Volume 1 Compilation
The Yoga Discovery Series:
Yoga-An Ancient Art Form
Hatha Yoga-Helping you Live Better
Raja Yoga-Through the Ages
The Yoga Discovery Package

Business & Coaching Books:
Creating, Publishing, & Marketing Practitioner Ebooks
Building a Successful Longevity Coaching Business
Why Become a Coach?
The Professional Coaching Success Trilogy
2020-Make Money Writing and Selling Books
The 2020 Handbook of High Paying Work Without a College Degree

Science, Technology, and Misc.
Future Predictions By and Engineer & Seer
The Unusual Science & Technology Bundle
The Real Atlantis-In the Eye of the Sahara
Are Cryptozoological Animals Real or Imaginary?
Real Time Travel Stories From a Psychic Engineer
Removing Limits On Our Consciousness-And Thinking Outside the Box
33 Incredible True Survival Stories
How to Survive Anything: From the Wilderness to Man Made Disasters
All About Mars Journeys and Settlement
Mining the Asteroid Belt

Ancient History
The Real Atlantis-In the Eye of the Sahara
Ancient & Prehistoric Civilizations
Ancient & Prehistoric Civilizations-Book Two
The History of Antediluvian Giants
The Antediluvian History of Earth
Ancient Underground Cities and Tunnels
Strange Objects Which Should Not Exist
Strange and Ancient Places in the USA
A Theory of Ancient Prehistory And Giant Aliens

Aliens and Space
Aliens and Secret Technology
Aliens Are Already Among Us
Designing and Building Space Colonies
Humanity and the Universe
All About Moon Bases
All About Mars Journeys and Settlement
The Space and Aliens Six Books Bundle
A Theory of Ancient Prehistory and Giant Aliens

| | |
|---|---|
| The Space Colonies and Space Structures Coloring Book | All About Asteroids |

The Longevity Training Series

*(A transcription of the online Multimedia Longevity Coaching Training Program)*

The Personal Longevity Training Series-Book1-Long Lived Persons
The Personal Longevity Training Series-Book2-Your Soul's Purpose
The Personal Longevity Training Series-Book3-Enable Your Life Urge
The Personal Longevity Training Series-Book4-Your Spiritual Connection
The Personal Longevity Training Series-Book5-Having Love in Your Heart
The Personal Longevity Training Series-Book6-Energy Body Health
The Personal Longevity Training Series-Book7-The Science of Longevity
The Personal Longevity Training Series-Book8-Physical Body Health
The Personal Longevity Training Series-Book9-Avoiding Accidents
The Personal Longevity Training Series-Book10-Implementing These Principles

The Personal Longevity Training Series-Books One Thru Ten

These books are all available in digital and printed formats from my website and on Amazon, Barnes & Noble, Apple ITunes, and many other sites

My Books Website is: http://mkettingtonbooks.com

## Signup for our Mailing List to get the following:

1) A discount coupon for 25% discount on all books on our site

2) Occasional Notices of new books available

3) Occasional Email on other offerings of ours (Monthly)

Go to this link to sign-up:

http://personal-longevity.com/mkebooks/emailsignup/

And click this link to get the FREE 102 page Ebook titled "Secrets of Many Things"

If you have any questions about this book or other subjects please contact the Author at:

mke@mkettingtonbooks.com

Aliens and Secret Technology

Aliens and Secret Technology

## Table of Contents

Chapter 1: Introduction .................................................................................. 10

Chapter 2: The Probability of Aliens Existing ............................................... 13

    a. The Drake equation ............................................................................. 13

    b. What Astronomy and Astrophysics tells us .......................................... 13

    c. The possibility of Advanced Alien Life ................................................. 14

Chapter 3: How to determine the Truth ........................................................ 17

    a. Evidence and Validity .......................................................................... 17

    b. Public Data Analysis ........................................................................... 17

    c. Objective analysis of Subjective Reports ............................................ 18

    d. Scale of Believability ........................................................................... 18

Chapter 4: Subjects I'm avoiding .................................................................. 23

    a. Crop Circles ........................................................................................ 23

    b. Alien related paranormal experiences ................................................ 23

    b. Alternate dimensions or realities ......................................................... 23

    c. Aliens on the Astral Plane ................................................................... 24

    d. Alien Abductions ................................................................................. 25

Chapter 5: The History of UFO & Aliens ....................................................... 27

    a. Ancient grooved spheres .................................................................... 27

    b.    Legends of the Annunaki ................................................................ 28

    c. Paintings on Stone walls in Northern Italy ........................................... 30

    d. The story of Ezekiel and the flying machine ....................................... 30

    e. Details of a fresco entitled "The Crucifixion" ....................................... 32

    f. The Dogon Tribe in Africa .................................................................... 33

    g. UFO Sightings by the Puritans ............................................................ 35

    h. Sightings in the Nineteenth Century ................................................... 38

Chapter 6: The Nazi Bell ...................................................................... 41

Chapter 7: Electro Gravitic Technology ........................................... 45

Chapter 8: Roswell New Mexico ...................................................... 47

Chapter 9: Alien Technology Transfer ............................................ 49

Chapter 10: Alien Bases on Earth ................................................... 53

Chapter 11: Zero Point Energy ........................................................ 55

Chapter 12: Serpo—Missions to Other Planets ............................ 57

Chapter 13: Caret Anti-Gravity Technology .................................. 61

Chapter 14: UFOS & Aliens Visitors ............................................... 65

Chapter 15: Types of Alien Races ................................................... 69

Chapter 16: Types of Likely Secret Craft ....................................... 71

   a. The Pumpkin Seed Hypersonic Craft ...................................... 71

   b. The TR 3B ................................................................................... 72

   c. Anti-Gravity and Element 115 .................................................. 73

d. Anti-Gravity air/space craft list ..................................................... 74

Chapter 17: The Secret Space Force ............................................. 83

Chapter 18: The United States incredible technology ................. 89

Chapter 19: Summary-My Theory of aliens and secret technology .......... 91

Bibliography ......................................................................................... 93

Index ..................................................................................................... 96

# Aliens and Secret Technology

## Chapter 1: Introduction

I have always been interested in Aliens and Secret Technology but hadn't really spent any time digging into the subject until the year 2000. Previous to 2000 I had spent many years studying the metaphysical, paranormal, spiritual, and occult, so didn't seem to have a lot of extra attention available to get into a whole new area of mysteries.

My background is an engineer and IT consultant so I have a pretty analytical view of the world. I've felt that this background gives me an advantage in researching controversial subjects and being able to sort out the facts from the fiction. I've also done work for some large Aerospace firms including a classified project for Boeing on the Ground Based Missile Defense from 1998 to mid-2004. (Although I admit never seeing any classified information on these subjects)

As I started to read books, websites, and blogs about UFOs and classified aerospace technology my fascination grew and I realized that there must be some truth to what people were reporting. The problem is that there is so much material available now that much of it is contradictory and maybe a portion is disinformation by the government.

My effort in this book is to provide some sense of the scope of potential alien influence in world history, and how may have influenced modern technology including secret aerospace technology. The objective is to provide an integrated theory of what is likely going on in our world on these subjects which makes some sense based on the facts.

I have tried to cover just the major events in history regarding aliens and secret technology. There are too many sightings, books, and videos to cover even a small portion of them.

This book is also not intended as an in depth description of many of the subjects and cases I describe. I'm merely providing an overview to provide context for my theories. If you want the details of each subject then please see the references in the Appendix

Hope you enjoy the journey.

Aliens and Secret Technology

# Aliens and Secret Technology

## Chapter 2: The Probability of Aliens Existing

In this chapter we explore what science and astronomy can tell us about the likelihood of alien races in our galaxy. (The Milky Way)

### a. The Drake equation

Dr. Frank Drake an astronomer proposed an equation on the possibility of intelligent life at the first scientific meeting on the search for extraterrestrial life in 1960. (SETI). This equation has since become the definitive calculation for the existence of alien life.

The Drake equation (1) is shown below with definitions of its variables.

$$N = R^* \, f_p \, n_e \, f_l \, f_i \, f_c \, L$$

- $N$ = The number of communicative civilizations
- $R^*$ = The rate of formation of suitable stars (stars such as our Sun)
- $f_p$ = The fraction of those stars with planets. (Current evidence indicates that planetary systems may be common for stars like the Sun.)
- $n_e$ = The number of Earth-like worlds per planetary system
- $f_l$ = The fraction of those Earth-like planets where life actually develops
- $f_i$ = The fraction of life sites where intelligence develops
- $f_c$ = The fraction of communicative planets (those on which electromagnetic communications technology develops)
- $L$ = The "lifetime" of communicating civilizations

Once you fill in the variables this equation will tell give you an estimate of the number of civilizations on other planets that we should be able to communicate.

Now that we are finding lots of planets and expect to find many planets that can support life, the probability of intelligent life has gone up significantly.

When filling in the variables based on conservative assumptions and our current knowledge of others stars, we find that there should be thousands of intelligent civilizations in just our own galaxy to talk to.

### b. What Astronomy and Astrophysics tells us

A good place to start is with what our current understanding of science and astronomical observations of the Universe tell us is possible.

Most Astronomers calculate that the age of our Universe is between 13.5 to 14 billion years old. (2) That there are also galaxies and stars that were formed only a billion years after the big bang that started our universe.

Life on earth took several billion years to start and our planet is estimated to be 4.5 billion years old. (3) Therefore, if life exists on any planets in the universe it may have started as much as nine to ten billion years old.

In the 1990s our telescopes and astronomy tools also became good enough to start finding planets around other stars. As of 2017, 4,500 candidate exoplanets have been found around stars by the Kepler Observatory. Of these about ten are Earth like planets in that they are in the stars habitable zone and are rocky planets like Earth. Almost every star examined seems to have planets.

We also know that there are at least 100 billion stars in our galaxy, and maybe more. (There are an estimated two trillion galaxies overall.) Let's also make an assumption that there might be one out of one thousand Earth like worlds which have intelligent life. Given the above information collected from Kepler and other observatories we can conclude the following:

That there should be over 100 million planets in our Galaxy alone which support intelligent life. That is a lot of potential alien civilizations.


c. The possibility of Advanced Alien Life

If life existed or exists that started in the first galaxies then what would it be like today?
We may speculate that any life that old would have evolved to be as Gods to us or as advanced compared to us as we are to an Amoeba.

Even life on other planets that evolved in a similar timeframe to ours could be thousands or even millions of years more advanced.

What types of technologies and abilities would beings even thousands of years older than us possess?

Just looking at the advancement of technology over time we should expect Aliens to be able to do things which would seem as magic to us.

Take stealth aircraft for example: Today scientists are already experimenting on technologies to bend light around materials to provide full invisibility. This should be attainable in a few more years. (4)

Of course I'm also assuming that very advanced Aliens would be able to find a way around the relativistic light speed barrier to travel easily between starts.

Aliens and Secret Technology

In terms of their ability to hide themselves, they would have stealth capabilities for their ships and individuals that would include the ability to hide from us completely if they so desired. Here is a picture taken in Chile as part of a video which the photographer says in undoctored and might be an alien who thought he was invisible: (5)

I've studied and experience a lot of paranormal events in my life so I'm a strong believer in intelligent beings having these abilities too. In terms of mental and paranormal abilities I would also speculate that Aliens would be well advanced beyond ninety nine percent of humans alive today too.

If Aliens are so far advanced of us mentally and in paranormal abilities they might also be able to make us see illusions and control us with their minds if they were here on earth. The possibilities are endless.

The main reason to discuss all these possibilities is to get the reader to understand that true Alien life may be so far beyond us that we could not even comprehend their abilities.

Therefore, we shouldn't summarily dismiss even the craziest reports from people who report UFO and Alien experiences without a proper investigation.

## Chapter 3: How to determine the Truth

There are thousands of books and websites on UFOs and Aliens so I'm not going to try to produce any new information or compete with those books.

What is apparent from all the literature out there is that there are so many claims and theories about what is going on that some of them are contradictory with each other.

How then to know what is the good information, and what are personal fantasies or government disinformation? Here we present several methodologies to try to help the reader analyze the validity of UFO and Alien reports.

This should help improve the validity of the information presented and allow us to reach some more analytical conclusions about what is really true. I'm not going to bore you by using all of these techniques but you might want to apply them to your own examination of source documents and videos.

### a. Evidence and Validity

Physical evidence of an event is the best thing to have to prove a case. To validate evidence we need to examine the possible chain of custody, and whether there are multiple sources who report the same things regarding the evidence.

### b. Public Data Analysis

The first approach is one common to intelligence agencies throughout the world.
This involves collecting information from many public sources which is also known in intelligence circles as open source intelligence. (OSINT) (6)

Sources for OSINT can include media, web site, blogs, observation and reporting, and any other sources publicly available.

In OSINT the chief difficulty is in identifying relevant, reliable sources from the vast amount of publicly available information.

Once the information is gathered it needs to be analyzed to build a coherent picture of a story, and evaluated for consistency.

For example, multiple sightings of a UFO should be compared to look for similarities in the description of the sighting. The more common items described by multiple witnesses, the more sure one can be that the object sighted probably had those characteristics.

Similarly, when multiple witnesses describe the same situation this information can be used to fill in the puzzle pieces to get a clearer picture. An example of this might be what different persons describe about what is going on in Area 51 in Nevada.

c. Objective analysis of Subjective Reports

One of the concepts I described in my book "Using the Scientific Method to Study the Paranormal" is the idea of generating more objective observations from multiple subjective sources. One of the reasons a lot of debunkers use to justify their statements that UFO and Alien sightings are invalid is that the observers may have mental problems, are drugged, or just dreaming or imagining things.

A way to improve the validity of these observations is to build a matrix of personal information, and then commonality of observation statements to see which ones are consistent and how those claims map to individual peculiarities. The more observations we can collect to build our data pool the better.

d. Scale of Believability

In my same book "Using the Scientific Method to Study the Paranormal" I also reviewed the concept of the Scale of Believability.

When I was in my second year as an undergraduate engineering student, I had an opportunity to run a psychic research course during the 1975 January term at Rensselaer Polytechnic Institute, Troy, NY under the auspices of the Chairman of the Physics department. (He scheduled the course and I worked with him to develop the material and taught most of it.)

At that point I had also read a lot of books on metaphysics, and was a year into my own psychic development through meditation. I was the daily instructor to a class of about 25 other students, where we did experiments on everything from pyramids to trying to hear messages from the dead on recording tapes.

In looking for a way to explain psychic phenomena to these students I came up with my scale of believability which I have become an ever-stronger adherent to as I've gotten older.

The value of this scale is to establish what types of events are more believable than others. As you educate yourself and form opinions on whether you believe that a certain type of event happens you take the next step on the scale to believing in events which sound more outrageous.

At some point in your explorations and discovery you either move your opinions further to the right towards the unbelievable, or decide you have crossed the line into fantasy and stop.

Also, as new evidence is collected we can reasonably move a type of event along the Scale to show where its current believability lies.

The figure below shows what the Scale looks like:


The concept is simple:

Imagine a scale with one end called "Probably True", and the other end called "Impossible/Ridiculous"

Then start listing unusual types of phenomena along the scale as to

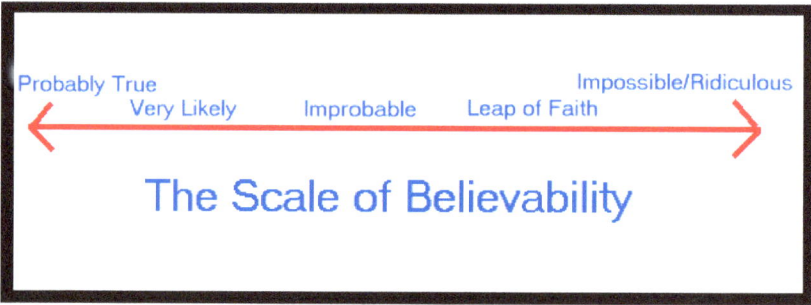

where they fall.

An example scale might include the following key, with these sample entries:

Probably True
Very Likely
Improbable
Leap of Faith
Impossible/Ridiculous

# Aliens and Secret Technology

Examples of points on the scale follow:

## UFOs Sighted—Probably True

With the thousands of sightings over more than 50 years, and some of them unexplainable, most people would say that really might be Alien ships from another planet.

## UFOs acquired by government-Very Likely

There have been enough stories about the Roswell and other spacecraft over the years that it isn't hard to believe that our government has already acquired Alien spacecraft

## Alien Abductions-Improbable

People being abducted by Aliens starts to sound pretty ridiculous. However, there are hundreds of reports of this happening.

## Many alien Races exist and have been to Earth-Leap Of Faith

To believe that there are many Alien Races, and that they have been to Earth and some are living here sounds pretty far out. However, there may be evidence to support this claim too.

## Aliens have Bases here on Earth and Live Among Us—Impossible

Most people would consider this ridiculous and impossible, but is it?

Many more items can be added to this scale. I just wanted to use some sample ideas to explain the concept.

Aliens and Secret Technology

Aliens and Secret Technology

## Chapter 4: Subjects I'm avoiding

Before we dig into the actual stories and evidence in this book I've also chosen what subjects should be part of the analysis and what to leave out. The following subjects are being left out due to the size of the whole subject of UFOs, Aliens, and Secret technology. There is too much to cover in any one book, so I want to just focus on the pieces which can fit together.

a. Crop Circles

Crop circles are a recent phenomena which have received a lot of press. It is a whole subject in itself and in my view is only tangentially related to UFOs and Aliens. There is so much to cover in the core areas that I'm leaving this one out.

b. Alien related paranormal experiences

It is possible to believe that advanced Aliens may have psychic powers and may be using those abilities to communicate with or control humans. However, these experiences tend to be reported as totally different by everyone and vary from some type of telepathy to meetings on the Astral Plane. I'm not saying that these events are imaginary, but they do not lend themselves well to any type of analytical analysis.

b. Alternate dimensions or realities

While Alternate dimensions or realities may exist in a scientific sense for proof of concepts of the structure of reality, there is no evidence that any intelligent beings exist in those other dimensions.

In my mind, it's also just too convenient to ascribe Alien beings as coming from another dimension when there isn't much in the way of real reports to support this. Therefore, I'm not covering these types of reports in my evidence analysis either.

c. Aliens on the Astral Plane

It is very probable that if there are aliens they would be much more advanced than us spiritually as well as technologically. They may well exist in non-physical or incorporeal forms. For purposes of this book however, I want to stick to our physical existence and what has been reported.

Aliens and Secret Technology

d. Alien Abductions

People being abducted by UFOs or aliens is a popular subject in stories about human interactions with UFO and aliens. I'm not covering these stories here because whether true or not, these stories are tangential to the information I'm building about technologies.

Aliens and Secret Technology

Aliens and Secret Technology

Chapter 5: The History of UFO & Aliens

There are many historical records that UFOs and Aliens have been observed in recorded history. There is also lots of evidence of anomalous technology that humans couldn't have created. Some of the best known include:

a. Ancient grooved spheres

Over the last few decades, miners in South Africa have been digging up mysterious metal spheres. Origin unknown, these spheres measure approximately an inch or so in diameter, and some are etched with three parallel grooves running around the equator. Two types of spheres have been found: one is composed of a solid bluish metal with flecks of white; the other is hollowed out and filled with a spongy white substance. The kicker is that the rock in which they were found is Precambrian - and dated to 2.8 *billion* years old!

Who made them and for what purpose is unknown. Something this old would have to have been created before any life existed on earth.

b. Legends of the Annunaki

There are lots books written on the subject of the Anunnaki who many of the Authors think were extraterrestrials meeting and affecting human civilization thousands of years ago.

Originally, the Anunnaki appear to have been heavenly deities with immense powers. In *Enki and the World Order*, the Anunnaki "do homage" to Enki, sing hymns of praise in his honor, and "take up their dwellings" among the people of Sumer. The same composition repeatedly states that the Anunnaki "decree the fates of mankind"

The ancient Sumerians claimed they were aliens who had ruled Earth for many centuries. Here is a short summary what the Sumerians believed:

> *The Sumerians possessed advanced understanding of mathematics, language and astronomy. However, there is no evidence of this knowledge evolving over time as one would expect. It just exists and is written about as being brought by the Anuna gods who came from the stars. The Sumerians are credited with establishing our modern day calendar and knew the cycles of the sun, moon, and visible planets. They knew the accurate length of a year.*
>
> *The Sumerians were the first civilization to divide space and time by factors of six.*

*The contemporary division of the year into 12 months, 24 hours days, hours into 60 minutes with 60 seconds are all documented by the Sumerians. This division by factors of six was later found at several other megalithic structures on other continents.*

*Sumerian understanding of time gives perspective to the writings known as the "Kings List." This chronicle of rule begins with original Anuna supreme rulers' decent from the heavens in the "kingship". This is said to have happened after a great flood. This is believed to be referring to the Biblical flood. It then lists all the Anuna rulers that followed and how long they ruled. Amazingly, many of them are listed to rule for hundreds of years. Other for much shorter times (30 years on average.) The first ruler held power for 1200 years! We cannot imagine this to be true since our human life span is 70-90 years on average. However, consider that a highly advanced being, millions of years more developed than we are, may have achieved this kind of longevity. Consider our own life span average of 70-90 years compared to that of humans during the first century BC that lived 30 years on average. Now, imagine what effect a million years of evolution may have on our life span.*

*The Kings List chronicles 433,000 years of Anuna rule. This means the age of the Anuna gods spanned a much longer period of time than the recorded history of all civilizations on Earth combined. The Anuna/Annunaki would have been on earth 430,000 years before any evidence of recorded history that has been discovered. We cannot imagine this amount of time since our limited experience in recorded history is less than 6,000 years in total (from Mesopotamia to today.) The accounts of people living these lengths of time are also found in the Hebrew Torah and Christian Bible's texts called the Old Testament. The real question is just how old are these accounts of time? Likely much older than any of us can possibly imagine.*

c. Paintings on Stone walls in Northern Italy

Paintings on stonewalls in Val Camonica, North Italy, showing beings seemingly wearing a glasslike helmet from which short rays are emitted. Dated 4,000 B.C. These could be some beings in spacesuits

d. The story of Ezekiel and the flying machine

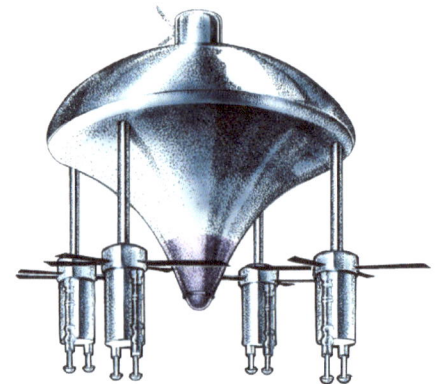

In about 600 B.C Ezekiel made one of the earliest accounts of some type of flying machine. It is recorded in the Old Testament of the Bible and here are some of the passages:

> And I looked, and, behold, a whirlwind came out of the north, a great cloud, and a fire infolding itself, and a brightness was about it, and out of the midst thereof as the color of amber, out of the midst of the fire. (Ezek. 1:4)

> And I looked, and, behold, a whirlwind came out of the north, a glowing cloud, and brilliant fire flashing itself in a circle; and in the

*midst thereof, an appearance of polished metal (or, gleaming electrum), in the midst of the fire.*

*As for the likeness of the living creatures, their appearance was like burning coals of fire, and like the appearance of lamps: it went up and down among the living creatures. (Ezek 1:13)*

This sighting also seems to indicate that he was abducted as he later reports rising into the air.

Aliens and Secret Technology

e. Details of a fresco entitled "The Crucifixion"

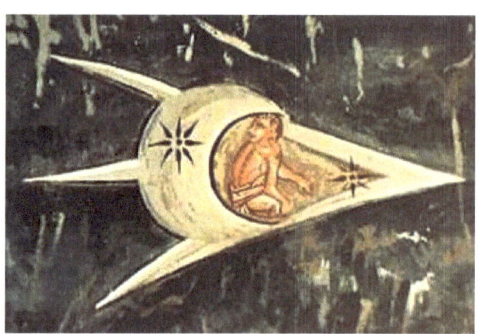

Details of a fresco entitled "The Crucifixion". They are located above the altar at the Visoki Decani Monastery in Kosovo. The year it was painted was 1380 A.D.

Diego Cuoghi identified the UFO: it looks just like symbolic and personified representations of the sun and the moon in numerous other paintings of the time.

Aliens and Secret Technology

f. The Dogon Tribe in Africa

In Mali, West Africa, lives a tribe of people called the Dogon. (7) The Dogon are believed to be of Egyptian decent and their astronomical lore goes back thousands of years to 3,200 BC. According to their traditions, the star Sirius has a companion star which is invisible to the human eye. This companion star has a 50 year elliptical orbit around the visible Sirius and is extremely heavy. It also rotates on its axis.

This legend might be of little interest to anybody but the two French anthropologists, Marcel Griaule and Germain Dieterlen, who recorded it from four Dogon priests in the 1930's. Of little interest except that it is exactly true.

How did a people who lacked any kind of astronomical devices know so much about an invisible star? The star, which scientists call Sirius B, wasn't even photographed until it was done by a large telescope in 1970.

The Dogon stories explain that also. According to their oral traditions, a race people from the Sirius system called the Nommos visited Earth thousands of years ago. They also appear in Babylonian, Accadian, and Sumerian myths. The Egyptian Goddess Isis, who is sometimes depicted as a mermaid, is also linked with the star Sirius.

The Nommos, according to the Dogon legend, lived on a planet that orbits another star in the Sirius system. They landed on Earth in an "ark" that made a spinning decent to the ground with great noise and wind. It was the Nommos that gave the Dogon the knowledge about Sirius B.

It doesn't seem to explain a 400-year old Dogon artifact that apparently depicts the Sirius configuration nor the ceremonies held by the Dogon since the 13th century to celebrate the cycle of Sirius A and B. It also doesn't explain how the Dogons knew about the super-density of Sirius B, a fact only discovered a few years before the anthropologists recorded the Dogon stories.

The Dogons are a people well known by their cosmogony, their estericism, their myths and legends that interest foreigners at the highest point in search for culture or tourism. The population is assessed to be about 300,000 people living in the South West of the Niger loop in the region of Mopti in Mali (Bandiagara, Koro, Banka), near Douentza and part of the North of Burkina (North west of Ouahigouya).

The Dogon's (Mali, Africa) homeland has been designated a World Heritage site for its cultural and natural significance. They are also famous for their artistic abilities and vast knowledge about astrology, especially the Sirius star, which is the center of their religious teachings. The Dogons know that Sirius A, the brightest system in our firmament, is next to a small white dwarf called Sirius B, which was not identified by western scientists until 1978. The Dogons knew about it at least 1000 years ago. Sirius B has formed the basis of the holiest Dogon beliefs since antiquity.

Western astronomers did not discover the star until the middle of the nineteenth century, and it wasn't even photographed until 1970. The Dogons go as far as describing a third star in the Sirius system, called "Emme Ya" that, to date, has not been identified by astronomers. In addition to their knowledge of Sirius B, the Dogon mythology includes Saturn's rings and Jupiter's four major moons. They have four calendars, for the Sun, Moon, Sirius, and Venus, and have long known that planets orbit the sun.

### g. UFO Sightings by the Puritans

During the 1600s, Puritans in New England spotted more than just witches flying through the skies. Hundreds of years before Area 51 and Project Blue Book, Massachusetts Bay Colony founder John Winthrop detailed instances of unidentified flying objects in the heavens above seventeenth-century Boston in the first recorded UFO sightings in America.

On March 1, 1639, John Winthrop opened his diary in which he recorded the trials and triumphs of his fellow Puritans as they made a new life in a new land. As the governor of the Massachusetts Bay Colony put pen to paper, he began to recount a most unusual event that had recently caused a stir among the English immigrants.

# Aliens and Secret Technology

Portrait of John Winthrop

*Winthrop wrote that earlier in the year James Everell, "a sober, discreet man," and two others had been rowing a boat in the Muddy River, which flowed through swampland and emptied into a tidal basin in the Charles River, when they saw a great light in the nighttime sky. "When it stood still, it flamed up, and was about three yards square," the governor reported, "when it ran, it was contracted into the figure of a swine." Over the course of two to three hours, the boatmen said that the mysterious light "ran as swift as an arrow" darting back and forth between them and the village of Charlestown, a distance of approximately two miles. "Diverse other credible persons saw the same light, after, about the same place," Winthrop added.*

*The governor wrote that when the strange apparition finally faded away, the three Puritans in the boat were stunned to find themselves one mile upstream—as if the light had transported them there. The men had no memory of their rowing against the tide, although it's possible they could have been carried by the wind or a reverse tidal flow. "The mysterious repositioning of the boat could suggest that they were unaware of part of their experience. Some researchers would interpret this as a possible alien abduction if it happened today," write Jacques Vallee and Chris Aubeck in "Wonders in the Sky: Unexplained Aerial Objects from Antiquity to Modern Times."*

*Some have speculated that the curious glow could have been an "ignis fatuus," a pale light that can appear over marshland at* nighttime due to the combustion of gas from decomposed organic matter. If Winthrop's report was correct, however, the light was not rising from the swamp but shooting across the sky, making that explanation unlikely.

An odd sight returned to the skies of Boston five years later, according to another entry in Winthrop's diary dated January 18, 1644. "About midnight, three men, coming in a boat to Boston, saw two lights arise out of the water near the north point of the town cove, in form like a man, and went at a small distance to the town, and so to the south point, and there vanished away."

A week later, Winthrop wrote, another unexplained celestial event occurred over Boston Harbor. "A light like the moon arose about the N.E. point in Boston, and met the former at Nottles Island, and there they closed in one, and then parted, and closed and parted diverse times, and so went over the hill in the island and vanished. Sometimes they shot out flames and sometimes sparkles. This was about eight of the clock in the evening, and was seen by many.

"About the same time a voice was heard upon the water between Boston and Dorchester, calling out in a most dreadful manner, 'Boy! Boy! Come away! Come away!'; and it suddenly shifted from one place to another a great distance, about twenty times. It was heard by diverse godly persons. About 14 days after, the same voice in the same dreadful manner was heard by others on the other side of the town towards Nottles Island."

Unlike the 1639 UFO, Winthrop had an explanation for the latest luminescence over his "city upon a hill." The governor noted the bizarre spectacle was seen near the location where a vessel captained by John Chaddock exploded months earlier after a sailor accidentally ignited gunpowder aboard the ship. The captain was not aboard at the time, but the blast killed five crew members.

Winthrop noted that rescuers had recovered the bodies of all the victims except for the man believed responsible for the calamity, a sailor who professed the ability to communicate with the dead and was suspected of murdering his master in Virginia. The hand of the devil was thought to have taken possession of the body, and it was the haunting voice of the sailor's ghost that was said to have accompanied the strange vision of Ye Olde UFO that mystified Boston.

## h. Sightings in the Nineteenth Century

There are also sightings in more recent history so these mysterious events aren't just limited to ancient reports. Tales of alien spaceships on Earth appeared in 19th Century papers:

> "About 35 miles northwest of Benkelman, Dundy County, on the 6th of June [1884] a very startling phenomenon occurred. It seems that John W. Ellis and three of his herdsmen and a number of other cowboys were out engaged in a round-up. They were startled by a terrific whirring noise over their heads, and turning their eyes saw a blazing body falling like a shot to earth. It struck beyond them, being hidden from view by a bank."

> The article, from the Nebraska Nugget, goes on to say that the rancher found "fragments of cog-wheels, and other pieces of machinery" lying on the ground. The heat was so intense that "as to scorch the grass for a long distance around each fragment and make it impossible for one to approach..." The group found the main part of the wreck and one of them "fell senseless from the gazing at it at too close quarters. His face was blistered, and his hair singed to a crisp."

> "Finding it impossible to approach the mysterious visitor [the UFO] the party turned back on its trail. When it [the UFO] first touched the earth the ground was sandy and bare of grass. The sand was fused to an unknown depth over a space about 20 feet wide by 30 feet long, and the melted stuff was still bubbling and hissing."

Later in the story the ship is described as being 50 to 60 feet long, cylindrical and 10 to 12 feet in diameter. The writer notes that it was apparently composed of metal with an appearance like brass, but was remarkably light. The story also notes that the wreck is located in a remote and wild region and "the roads are hardly more than trails."

The second story appeared in the *Dallas Morning News* on April 19th, 1897:

> "About 6 o'clock this morning the early risers of Aurora were astonished at the sudden appearance of the airship that had been sailing throughout the country. It was traveling due north, and much nearer the earth than before."
>
> The article describes how the air vehicle "sailed over the public square and when it reached the north part of town collided with the tower of Judge Proctor's windmill and went to pieces with a terrific explosion, scattering debris over several acres of ground, wrecking the windmill and water tank and destroying the judges' flower garden." It continues with, "The pilot of the ship is supposed to have been the only one aboard, and while his remains are badly disfigured, enough of the original has been picked up to show he was not an inhabitant of this world."
>
> "Mr. T. J Weems, the U.S. Signal Service officer at this place and an authority on astronomy, gives it as his opinion that he [the pilot] was a native of Mars." According to the story the remains of the ship were composed of a strange metal that seemed a mixture of aluminum and silver. The townspeople came to view the wreck and pick up specimens. The pilot was buried the day after the article was published.

There are many more examples of mysterious and anomalous evidence that some type of beings visited the earth in our past. There are enough examples here since there are many more books on historical UFO and Alien records.

My purpose here was just to show that this subject didn't just start recently but has been a controversial subject for all of human history.

Aliens and Secret Technology

Chapter 6: The Nazi Bell

Germany in World War Two had many secret projects which became a reality. The best known are the V1 and V2 buzz bombs and rockets which rained destruction on Britain throughout the war.

Other projects included secret work on a Nazi Atomic bomb. Although conventional wisdom is that the Nazi never came close to building a bomb, several recent books argue that they may have actually exploded a couple of test nuclear devices in late 1944 but were short of materials and delivery systems to ever turn this weapon into practical usage. (8)

Another project the Nazis developed and used in battle successfully was the first practical jet fighter. This was called the ME262. Fortunately, there weren't many of them because they managed to shoot down lots of allied bombers and fighters late in the war.

Additional research was conducted on what was called the Nazi Bell. This was the most classified project in Germany during World War Two. This project was worked on mainly in Poland and Czechoslovakia and involved a large bell shaped object with counter rotating cylinders. Extremely high voltage was injected into the assembly.

Once activated, the Bell had an "effect zone," or a zone in which certain things would happen, that stretched between 490 and 660 feet. Some claim that extremely strange things occurred within this zone, such as the formation of crystals within animal tissue, the decomposition of plant matter into a greasy substance, and the gelling and separation of blood.

The Bell was powered with a liquid fuel known as Serum 525 or Xerum 525. Proclaimed insiders describe it as cherry red, velvet or maroon in color and viscous, dense and gooey in texture. Some, like author Henry Stevens claim this liquid was actually red mercury.

There is a great book about the whole subject titled "SS Brotherhood of the Bell: The Nazis' Incredible Secret Technology". (9) This book gives a lot more details including that this experimental device likely had to do with anti-gravity.

There are two most interesting facts historically regarding the Nazi Bell. First is that the American Army occupied this area where research was being done first in the spring of 1945. The U.S. would have had time before handing it over to the Russians to remove documentation and any scientists to the U.S. It is well known that many V2 rocket scientists came to the U.S. after the war through "Operation Paperclip" including core scientists for the U.S. space program.

The second fact is that the general providing overall direction on this and other secret Nazi projects named "Hans Kammler" was never caught or found after the war. There are some stories that he too was brought to the U.S. for secret projects going on here.

So the Nazi bell might have been the foundation of United States research on anti-gravity in 1945.

Aliens and Secret Technology

Aliens and Secret Technology

## Chapter 7: Electro Gravitic Technology

Electrogravitics is claimed to be an unconventional type of effect or anti-gravity propulsion created by an electric field's effect on a mass. The name was coined in the 1920s by the discoverer of the effect, Thomas Townsend Brown, who spent most of his life trying to develop it and sell it as a propulsion system.

Through Brown's promotion of the idea it was researched for a short while by aerospace companies in the 1950s. Electrogravitics is popular with conspiracy theorists with claims that it is powering flying saucers and the B-2 Stealth Bomber.

Some people claim that the extremely high prices of the B-2 were because additional electrogravitic B-2s were produced for every one bomber seen by the public.

Since apparatus based on Browns' ideas have often yielded varying and highly controversial results when tested within controlled vacuum conditions, the effect observed has often been attributed to the ion drift or ion wind effect instead of anti-gravity. Brown also named this phenomenon the "Biefeld–Brown effect" after his claimed mentor, Denison University professor Paul Alfred Biefeld. **A picture of Thomas Townsend Brown is below:**

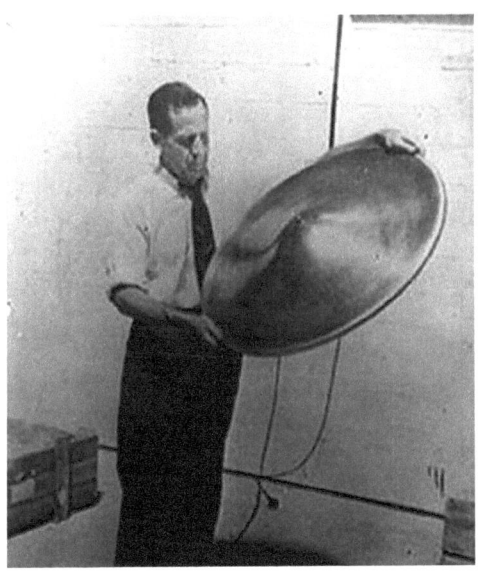

Aliens and Secret Technology

Aliens and Secret Technology

Chapter 8: Roswell New Mexico

Ahh… The Roswell Incident. How many thousands of articles and books have been written about the likely alien ship crash near Roswell, New Mexico in 1947. There is now even a museum there to visit.

Here is one famous newspaper issued after the crash before the military changed their story to say a weather balloon had crashed:

Most retellings of this story include that dead alien bodies were recovered. Some stories are that the aliens were alive and were kept in isolated conditions for years.

How to separate fact from fiction? I think it's fair to say that much of the UFO culture around the world is based on this incident. I have read several books which talked about testimonies of aliens captured from the accident, technology developed as a result, and more.

All I can say for sure is that this is the biggest event in recorded UFO history so it's likely there was some basis in fact. Other chapters in this book will discuss possible results of this crash on secret and advanced technology.

Aliens and Secret Technology

Aliens and Secret Technology

Chapter 9: Alien Technology Transfer

The book "The Day After Roswell" by Colonel Philip J. Corso is really interesting and says that we reversed engineered lots of alien technology after the Roswell crash. Having read the whole story I have to give it an "A" for consistency and it's a plausible story. There are a couple other stories like this I've seen but this one is the classic.

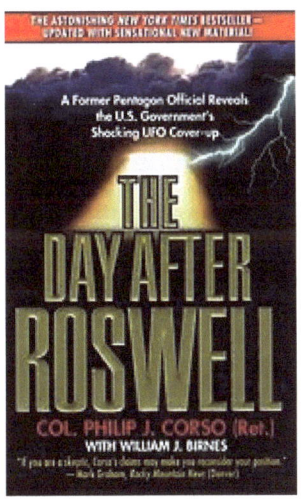

The claims in this book are that everything from night vision goggles to transistors, and fiber optics were reverse engineered and given to industry as a result of the Roswell crash.

That we also learned how to reverse engineer and build anti-gravity ships from it too.

I do question some claims such as those about transistors, because there is a lot of documentation that they were developed at Bell Labs in New Jersey by two scientists there and didn't just appear fully developed. But who knows? Transistors could have been planted.

Here are a few paragraphs from the book about development of the laser:

> The retrieval team that pulled the wreckage out of the desert also found a short, stubby, internally powered flashlight device that threw a pencil thin, intense beam of light for a short distance that could actually cut through metal. This, the engineers at Wright

*Field believed, was also based on wave stimulation. The questions then were, how did the EBEs use wave stimulation and how could we adapt it to military uses or slip it into the product development already under way?*

*By 1954, when I was at the White House, the National Security Council was already receiving reports of a theory, developed by Charles H. Townes, that described how the atoms of a gas could be excited to extraordinarily high energy levels by the application of bursts of energy. The gas would release its excess energy as microwaves of a very precise frequency that could be controlled. In theory, we thought, the energy beam could be a signal to carry communications or an amplifier for the signal. When the first maser was assembled at Bell Laboratories in 1956, it was used as a timer because of the very exact calibration of the wave frequency.*

*The maser, however, was only a forerunner of the product that was to come, the laser, which would revolutionize every aspect of technology it touched. It would also prove to be a weapon that would help us deploy a realistic threat to the EBEs who seemed poised to trigger a nuclear war between the superpowers. Where the maser was an amplification of generated microwaves, the laser was an amplification of light, and theories about how this might be accomplished were circulating widely throughout the weapons development community even before Bell Labs produced the first <u>maser</u>. I had seen the descriptions of the EBE laser in reports about the Roswell crash, a beam of light so thin that you couldn't even see it until it landed on a target.*

*What was the purpose of this light generator? the Alamogordo group had asked. It looked like a targeting or communications device, seemed to have an almost limitless range, and, if the right power source could be found to amplify the light beam to where it could penetrate metal, the device could be used as a drill, a welder, or even a devastating weapon.*

*Even while I was at the White House, all three branches of the military were working with researchers in university laboratories to develop a working laser. In theory, exciting the atoms of an element to produce light energy in the same way that atoms of a gas were excited to produce microwaves, lasers offered the tantalizing promise of a directed energy beam that had such a wide variety of applications it could become an almost universal utility for all divisions of the military, even controlling warehouse inventory for the Quarter master Corps. Finally, in 1958, the year after I left the White House, there was a surge in research activity, especially at Columbia University where, two years later, physicist Theodore Maiman constructed the first working laser*

Aliens and Secret Technology

Aliens and Secret Technology

Chapter 10:  Alien Bases on Earth

The story of alien bases on Earth or under the Earth or even under the ocean is a whole area of discussion by itself. I almost decided to not include this in this discussions of alien technologies, except if true, the relationships with aliens in these bases would have a profound impact on our knowledge of alien technologies.

The greatest claim about an alien base is the one for the underground base at Dulce, New Mexico.

Starting in 1979, Bennewitz became convinced he was intercepting electronic communications from alien spacecraft and installations outside of Albuquerque. By the 1980s he believed he had discovered an underground base near Dulce. The story spread rapidly within the UFO community and by 1990, **UFOlogist** John Lear claimed he had independent confirmations of the base's existence.

Political scientist **Michael Barkun** writes that **Cold War** underground missile installations in the area gave superficial plausibility to the rumors, making the Dulce base story an "attractive legend" within Ufology. According to Barkun, claims about experiments on abductees and firefights between aliens and the **Delta Force** place the Dulce legend "well outside even the most far-fetched reports of secret underground bases."

Some claim that there is even an underground tunnel network connecting bases internationally (10):

> An underground Military Base/Laboratory in Dulce, New Mexico connects with the underground network of tunnels which honeycombs our planet, and the lower levels of this base are allegedly under the control of Inner Earth beings or Aliens. This base is connected to Los Alamos research facilities via an underground "tube-shuttle." (It can be assumed that such a shuttle way would be a straight-line construction. It should then be possible, by using maps and some deduction, to determine the most likely location of this base, especially since the general location is already known.) Beginning in 1947, a road was built near the Dulce Base, under the cover of a lumber company. No

*lumber was ever hauled, and the road was later destroyed. Navajo Dam is the Dulce Base's main source of power, though a second source is in El Vado (which is also another entrance). (Note: The above facts should also help to locate the base.) Most of the lakes near Dulce were made via government grants "for" the Indians. (Note: The September, 1983 issue of Omni (Pg. 80) has a color drawing of 'The Subterrene,' the Los Alamos nuclear-powered tunnel machine that burrows through the rock, deep underground, by heating whatever stone it encounters into molten rock, which cools after the Subterrene has moved on. The result is a tunnel with a smooth, glazing lining.)" (Note: Where would the molten rock go? And what has been done with this concept since 1983?)*Here is a diagram of possible underground tunnels connecting various secret bases together:

Aliens and Secret Technology

## Chapter 11: Zero Point Energy

Zero Point energy is real and is accepted by physicists. If we found a way to harness this energy it would power all our technology with no limitations. Here is a physics explanation of it:

> Zero-point energy (ZPE) or ground state energy is the lowest possible energy that a quantum mechanical system may have. Unlike in classical mechanics, quantum systems constantly fluctuate in their lowest energy state due to the Heisenberg uncertainty principle. As well as atoms and molecules, the empty space of the vacuum has these properties. According to Quantum Field Theory the universe can be thought of not as isolated particles but continuous fluctuating fields: matter fields, whose quanta are fermions (i.e. leptons and quarks), and force fields, whose quanta are bosons (e.g. photons and gluons). All these fields have zero-point energy. These fluctuating zero-point fields lead to a kind of reintroduction of an aether in physics, since some systems can detect the existence of this energy. However this aether cannot be thought of as a physical medium if it is to be Lorentz invariant such that there is no contradiction with Einstein's theory of special relativity.

In the popular book "The Hunt for Zero Point" Nick Cook (11) claims that much classified research is also about how to harness this energy for usage and that many of the experiments on the Nazi Bell among other projects had less to do with anti-gravity, and more to do with harnessing the unlimited potential of Zero Point Energy.

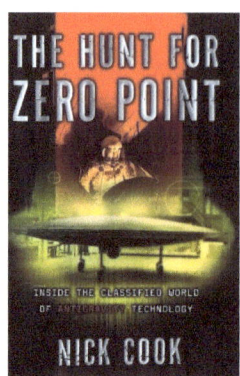

There are many conspiracy theories which says governments are suppressing this technology worldwide. I don't current have the answer to this question.

# Aliens and Secret Technology

## Chapter 12: Serpo—Missions to Other Planets

One of the most interesting stories in the whole UFO and Alien realm is the story of the Serpo Project. The Serpo project is about how twelve military personnel were selected for a mission to the planet of the Ebens and how they stayed for thirteen years before returning. (12)

For those of you who have seen the movie "Close Encounters of the Third Kind" it's also really interesting how the ideas of a team of military personnel traveling on an alien ship to another star system is a very parallel concept.

Here is the original message posted on serpo.org which gives basic information about the program:

> First let me introduce myself. My name is Request Anonymous. I am a retired employee of the U.S. Government. I won't go into any great details about my past, but I was involved in a special program.
>
> As for Roswell, it occurred, but not like the story books tell. There were two crash sites. One southwest of Corona, New Mexico and the second site at Pelona Peak, south of Datil, New Mexico.
>
> The crash involved two extraterrestrial aircraft. The Corona site was found a day later by an archaeology team. This team reported the crash site to the Lincoln County Sheriff's department. A deputy arrived the next day and summoned a state police officer. One live entity [EBE] was found hiding behind a rock. The entity was given water but declined food. The entity was later transferred to Los Alamos.
>
> The information eventually went to Roswell Army Air Field. The site was examined and all evidence was removed. The bodies were taken to Los Alamos National Laboratory because they had a freezing system that allowed the bodies to remain frozen for research. The craft was taken to Roswell and then onto Wright Field, Ohio.
>
> The second site was not discovered until August 1949 by two ranchers. They reported their findings several days later to the

sheriff of Catron County, New Mexico. Because of the remote location, it took the sheriff several days to make his way to the crash site. Once at the site, the sheriff took photographs and then drove back to Datil.

Sandia Army Base, Albuquerque, New Mexico was then notified. A recovery team from Sandia took custody of all evidence, including six bodies. The bodies were taken to Sandia Base, but later transferred to Los Alamos.

The live entity established communications with us and provided us with a location of his home planet. The entity remained alive until 1952, when he died. But before his death, he provided us with a full explanation of the items found inside the two crafts. One item was a communication device. The entity was allowed to make contact with his planet.

Somehow, I never knew this information, but a meeting date was set for April 1964 near Alamogordo New Mexico. The Aliens landed and retrieved the bodies of their dead comrades. Information was exchanged. Communication was in English. The aliens had a translation device.

In 1965, we had an exchange program with the aliens. We carefully selected 12 military personnel; ten men and two women. They were trained, vetted and carefully removed from the military system. The 12 were skilled in various specialties.

Near the northern part of the Nevada Test Site, the aliens landed and the 12 Americans left. One entity was left on Earth. The original plan was for our 12 people to stay 10 years and then return to Earth.

But something went wrong. The 12 remained until 1978, when they were returned to the same location in Nevada. Seven men and one woman returned. Two died on the alien's home planet. Four others decided to remain, according to the returnees. Of the eight that returned, all have died. The last survivor died in 2002.

[Clarification (BR/ Victor Martinez): the paragraph above contains a typo in the original. Twelve team members went, and eight returned – two having died on Serpo and two having chosen to *remain; these two were not ordered to return.]*

*The returnees were isolated from 1978 until 1984 at various military installations. The Air Force Office of Special Investigation (AFOSI) was responsible for their security and safety. AFOSI also conducted debriefing sessions with the returnees.*

*I have never seen or read anything about the exchange program. I once heard a little bit of information from Linda Howe, but she didn't have much information.*

*I've monitored your e-mails for about six months. I've read e-mails from you and others. But I've never seen nor heard the truth about the real Roswell incident or the exchange program.*

*I'd like to hear what others say about this.*

\*\*\*\*\*

Aliens and Secret Technology

Below is a picture of one of the alien spaceships:

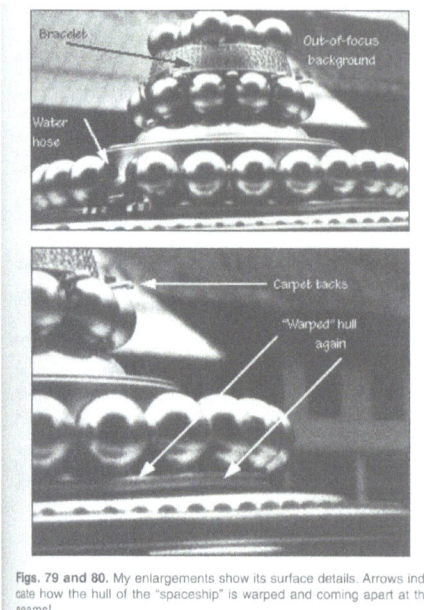

Figs. 79 and 80. My enlargements show its surface details. Arrows indicate how the hull of the "spaceship" is warped and coming apart at the seams!

Although all of the information sourced is from one website: www.serpo.org Another tie in is the internal consistency of the reports among themselves and the tie in to other Alien and UFO information. For instance, one the first alien abduction cases created a star map under hypnosis which showed the star Zeta Reticuli (Betty & Barney Hill- as discussed in the abduction chapter of their book) as where the aliens were from. This is the same star system that humans are supposed to have visited per Serpo.org.

Aliens and Secret Technology

Chapter 13: Caret Anti-Gravity Technology

In 2007 a strange program came to light which is called the "Caret" program or "The Palo Alto Caret Laboratory" and which is supposed to have been an alien technology re-engineering program based at a civilian company in Silicon Valley in the 1980s. The work was performed on a classified contract for the U.S. military. (13)

The story also includes eyewitnesses who saw a craft at different places in California which is supposed to be U.S. built and based on alien reverse engineered technology. A picture of the unmanned craft is below which has marking later described to be similar to those from the Caret program:

What makes this story so fascinating are pages from supposed secret documents which provide a lot of details about the technology. The program was about sophisticated alien materials which could be programmed to have different physical properties including anti-gravity.

The idea is that these Meta materials have microscopic designs which allow the different properties to manifest purely by drawing the correct diagram. The technology was like writing software to make a program run, but instead of a program the materials would actually become what is programmed.

If true, this technology is obviously hundreds if not thousands of years in advance of ours because it means building Nano-machines at an atomic level and those machines having the ability to manipulate gravity too.

Aliens and Secret Technology

Here is a picture of one of the programming diagrams:

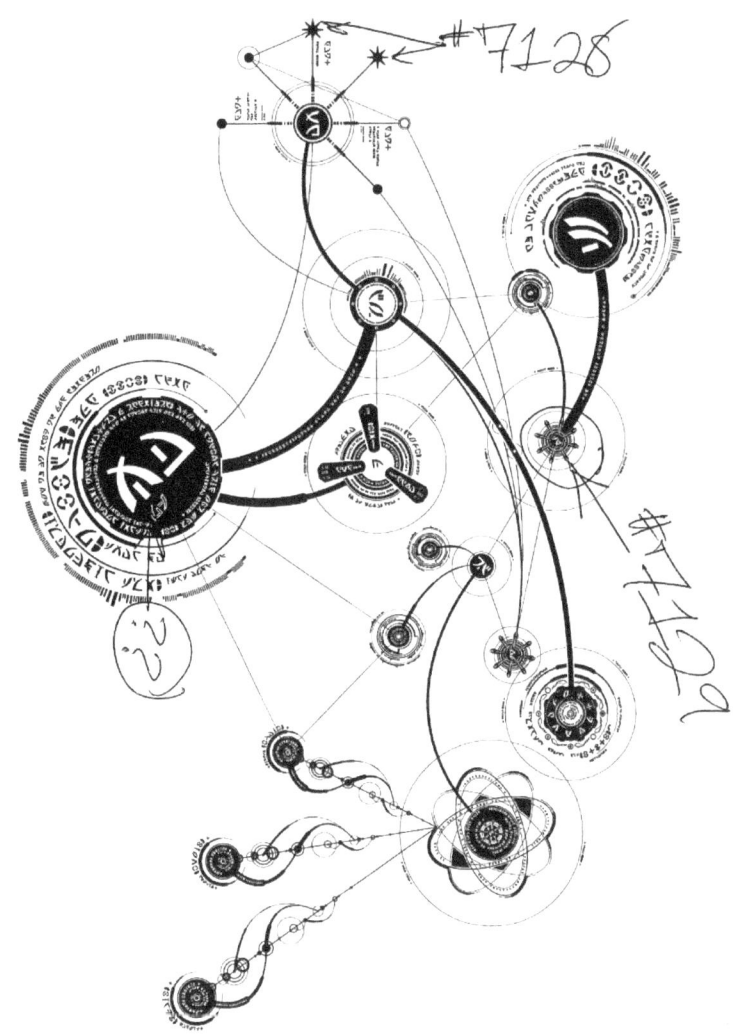

*Figure 14.11*
*Full view of diagram D39-08-117c.*

Aliens and Secret Technology

Another picture shows a Caret anti-gravity generator and two objects held in place by a designed anti-gravity field:

This program is so different from anything previously described in alien technology literature that I'm not sure what to make of it. The fact that unmanned craft were sighted numerous times by eyewitnesses makes me think this was a valid program.

# Aliens and Secret Technology

## Chapter 14: UFOS & Aliens Visitors

In 1961 there was apparently a UFO sighting in Europe which almost kicked off World War Three with the Soviets. This led to the NATO assessment document about the potential threats of multiple alien races.

Robert Dean, a retired United States army Sergeant Major says that he learned about this incredible NATO document back in the 1960s. He said the main document was only two inches thick but had appendixes eight inches thick documenting all of the claims in the main paper. Some of the aliens were little guys with green skin, some looked like they had vertical pupils and skin like a lizard.

Sergeant Dean also stated that there were twelve known alien species at the time he read the report but that since then from his government contacts he now believes there are as many as one hundred known alien species. That some of them look just like us and so you could be sitting next to one in a restaurant and would never know it.

In terms of verifying the validity of his statements Dean says "There have been other old boys like myself who have come forward and said ,'Dean is telling the truth, I was there, I read it,' including a retired Air Force three-star general and a couple of admirals. Recently a retired Soviet major-general (KGB), Oleg Kalugan, also confirmed the existence of the Assessment in a lecture in the States"

*****

I also read an interesting series of books a few years ago about a man in the Air Force in the 1960s who launched weather balloons at a remote base in the California desert.

He claims that the aliens have been using that area as a way station for alien travel for hundreds of years. He even met these aliens who he said could pass for humans with makeup.

One of the funny things he says is that the aliens like to hang out at the StarDust Hotel in Las Vegas because of the starry motif.

The book series is titled "Millennial Hospitality" By Charles James Hall (14)

*****

Aliens and Secret Technology

One of the most important books on aliens in recent years is "Disclosure" by Steven M. Greer M.D. It documents many very reputable alien and UFO cases around the world documented by military and other reputable witnesses. (15)

This book is probably the best documented and most credible book I've ever read which talks about UFOs and aliens visiting our planet.

So what are the overall probabilities that aliens have visited Earth? From the evidence I'd put that percentage at 99%.

Aliens and Secret Technology

Aliens and Secret Technology

Chapter 15: Types of Alien Races

In the previous chapter we referred to as many as 100 aliens already identified by governments around the world. Here are some of the most commonly identified alien races:

| Name | Alien Race Description | Image or Drawing |
|---|---|---|
| The Greys | Grey-skinned humanoids, usually 3–4 feet tall, bald, with black almond-shaped eyes, nostrils without a nose, slits for mouths, no ears and 3-4 fingers including thumb. They have been the centre of quite a few cases of alleged alien contact over the years. Most common type described who were in the Roswell crash. | |
| The Nordics | Humanoids with stereotypical "Nordic features" (tall, blond hair, blue eyes) and have featured in several cases of contact. It is said they are from Ancient Earth but presented themselves as ET's in the past, they moved from living in the surface to live underground around the Himalayas area after a natural event. Also the possible alien race living in the desert of California | |
| Draconians or Reptilians | Tall, scaly humanoids. Often described as having a hidden agenda on earth and hiding or imitating certain humans in power | |

A comprehensive list of possible alien species can be found in the references (16)

There are many more potential alien races and lists exist on the internet. I don't know how many of them are real or just fantasies. I listed the three above because they are the most common ones mentioned in the literature.

Aliens and Secret Technology

Chapter 16: Types of Likely Secret Craft

a. The Pumpkin Seed Hypersonic Craft

The Pumpkin Seed is a craft I'm 99% sure exists. The first story I ever saw about it was by Aviation Week & Space technology in the 1980s. The story talked about a new unmanned strategic craft which could overfly the Soviet Union at over 100,000 feet of altitude and drop over one hundred nuclear warheads out of it over Soviet territory.

Since then I've seen stories of this craft verified in different ways. The best verification example is the supposed usage of a pulse detonation technology which cuts in at about Mach 2 from the ship which starts with standard jet engines. The pulse is generated by the high speed flow of air over the craft where the fuel is pushed into the Windstream at the juncture line you can see in the below diagram. The fuel then catches fire and provides an immense propulsion force which pushes the speed up to Mach 12 or more. Something like 12,000 miles per hour.

The surface of the ship uses ceramic tiles like the space shuttle to keep it from melting.

Another interesting attribute of this ship is that it leave a "doughnuts on a rope" contrail from the engine pulses. Here is a picture:

I've actually seen this contrail back in the early 1990s when I was skiing at Mammoth Mountain in northern California, so I know it exists.

The impression this hypersonic ship made on the Soviets is supposed to have been very depressing that they couldn't protect themselves from it. This is one of several reasons the Soviet Union fell.

b. The TR 3B

The TR-3B is a triangular aircraft with three propulsion rockets at each of the three corners. Many pictures and videos have been snapped of this craft around the world. It seems to be another top secret craft of the United States. I also put the existence of this craft at over 99% since it has been seen over so much of the world.

It was first mentioned by Edgar Fouche who has described everything which is known publically about this craft. He claims that there is a mercury vapor being accelerated and rotated in a torus circle inside the perimeter of the triangular craft which reduces gravity on it by over 90%. This allows it to achieve maneuvers we would call impossible such as shooting from rest into motion of several thousand miles per hour and making incredibly sharp turns. This would be possible for a craft with only 10% actual effective mass.

**USAF Top Secret Nuclear Powered Flying Triangle - The TR-3B**

### c. Anti-Gravity and Element 115

Robert Scott Lazar (born January 26, 1959) claims to have worked on reverse engineering extraterrestrial technology at a site called S-4, near the Area 51 test facility, and that the UFOs use gravity wave propulsion.

This is powered by the, at the time, undiscovered element 115. He further claims to have read US government briefing documents that describe alien involvement in human affairs over the past 10,000 years. He runs a scientific supply company. Universities from which he claimed to hold degrees show no record of him.

He has described the complete layout of the ship he worked on per the image below:

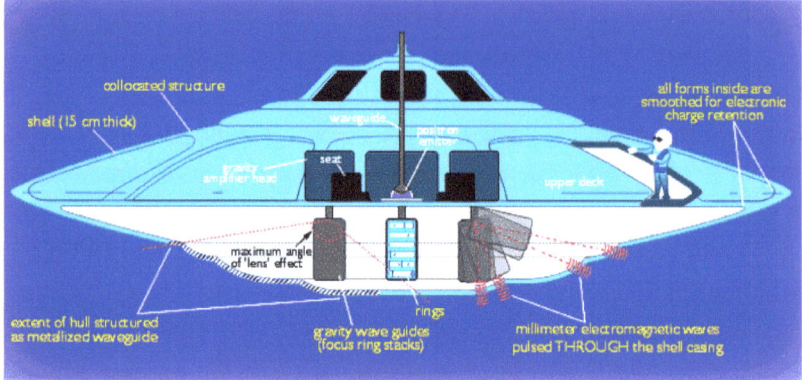

Because of the lack of other confirming evidence of his claims I have to put the probability of this being true at about 25%

d. Anti-Gravity air/space craft list

> Dr. Richard Boylan's list of twelve existing anti-gravity spacecraft. Dr. Boylan is a researcher on UFOs and aliens. (1)
>
> 1) The B-2 Stealth Bomber is manufactured by Northrop-Grumman. The Air Force describes it as a low-observable, strategic, long-range heavy bomber capable of penetrating sophisticated and dense air-defense shields. The Stealth Bomber has navigation and guidance systems directed by a classified Artificial Intelligence (AI) program. This AI is exotic, involving the connecting together PCR-cloned copies of extraterrestrial brain tissue with advanced integrated circuits to fashion hybrid sentient "neuro chips". The B-2 gets extra lift in-flight by the turning on of electro-gravitic fields along its wings and fuselage to considerably neutralize the pull of gravity. Now you begin to see why the U.S.'s Northrop B-2s cost about a billion dollars each.
>
> 2) The F-22 Raptor advanced Stealth Fighter is built by a joint effort of the Lockheed-Martin Skunk Works and Boeing's Phantom Works. In crude imitation of extraterrestrial starcraft, the guidance system of this F-22 aircraft incorporates special

*Artificial Intelligence (AI), meaning that cloned nerve pathways patterned on Star Visitor brain tissues are incorporated into the semi-sentient and semi-autonomously-functioning guidance system integrated with the aircraft. In addition, the F-22 has antigravity weight-reduction capability, which is exercised selectively by the pilot who is in mental connection with the AI guidance system, which activates the antigravity field generation as needed in coordination with the F-22's conventional jet engine thrust to effect the overall maneuvering of the F-22 Raptor. I have personally witnessed a F-22 pull up and hover vertically on its tail in mid-air, dance slowly, and gently rotate about on its vertical axis with little regard for the pull of gravity!*

3) *The "next generation" F-35 Lightning II advanced Stealth Fighter is built by Lockheed-Martin, Northrop-Grumman and BAE. The F-35 similarly incorporates quasi-sentient Artificial Intelligence and antigravity weight-reduction propulsion (back-engineered stolen Star Visitor technologies) in addition to its jet thrust.*

4) *The Aurora is a moderate-sized spacefaring vehicle. The late National Security Council scientist Dr. Michael Wolf(4) of MJ-12 [under cover as NSC's "Special Studies Group"] has stated that the Aurora can operate on both conventional-fuel and reduced gravity field-propulsion modes. Dr. Wolf further stated that the Aurora can travel to the Moon. Wolf had also disclosed to me that the U.S. has a small manned station on the Moon, and a tiny observation post on Mars. Thus I doubt that Dr. Wolf would characterize the Aurora as space-operational unless it was a vessel already used in making such trips. He disclosed additionally that the Aurora operates out of Area 51, (Groom Dry Lake Air Force Station), at the*

northeast corner of the Nellis AFB Range, north of Las Vegas, Nevada.

5) The Lockheed-Martin X-33A. Lockheed openly acknowledges its rocket-engine X-33 [no A] 'prototype', a single-stage-to-orbit reusable aerospace vehicle called the National Space Plane. Lockheed's other space plane, the X-33A, incorporates antigravity technology and is kept secret. Lockheed-Martin does not say too much about its conventional winged delta-shape X-33 VentureStar except to say that 'we are in the process of building it.' At that stage of development for its public-program Space Plane, it is clear that Lockheed-Martin has already long since secretly built its unacknowledged, advanced military operational antigravity version, which I have dubbed the X-33A. (The 'A' suffix stands for antigravity.) Colonel Donald Ware, USAF (ret.) told me that he had recently learned from a three-star General that the "VentureStar X-33" has an electrogravitics antigravity system on board(6.). He was really referring to the unacknowledged military antigravity version, the X-33 A, which indeed has electrogravitics on board.

6) The Lockheed X-22A is a two-man antigravity discoid craft. The late Colonel Steve Wilson, USAF stated that military astronauts trained at a secret aerospace academy adjacent to but separate and undergrounded from the regular Air Force Academy at Colorado Springs, CO. These military astronauts then operate out of Beale and Vandenberg Air Force Bases, Northern California. From those bases, these military astronauts regularly fly trans-atmospherically and out into deep space (7). The late Colonel Steve Wilson confirmed that one of the aerospace craft these astronauts use is the X-22A.

Another informant told me that the Lockheed X-22A antigravity disc is capable of achieving optical as well as radar invisibility.

*Recently I also heard from an Army engineer, formerly TDY'ed to NASA, who shall remain unnamed at his request. He also confirmed that Lockheed had made the X-22A, the two-man antigravity disc which I had seen test-flown in a canyon adjacent to the main Area 51 operations zone. He explained why I had seen the X-22A so gingerly flown during that test flight. He said that the original X-22A had had a standard altimeter hard-wired into it, but that such an instrument would give faulty readings in the craft's antigravity field (which bends space-time.) He had recommended that they instead use a gradiometer, which would function better. Apparently his suggestion was finally taken up, since in more recent years I have seen the X-22As flown more smoothly and confidently at high altitudes over and near Area 51.*

7) *The Nautilus is another space-faring craft, a secret military vehicle which operates by magnetic pulsing. It operates clandestinely out of Vandenberg Air Force Base, California. It makes trips several times a week up to the joint-nations secret military-intelligence space station. [This undeclared and cloaked space station has been in deep space for the past forty years, manned by U.S., Russian, British, Canadian, Austrian, Australian, Brazilian, and other military astronauts.] The Nautilus also is used for superfast surveillance operations utilizing its ability to penetrate target-country airspace from above entering from deep space, a direction not usually expected. It is manufactured jointly by Boeing's Phantom Works (Nellis Air Force Range, Nevada) and the Airbus Industries Anglo-French consortium. During travel to Washington State several years ago, I had a conversation with a former Boeing executive who worked in Boeing's Phantom Works black projects division, (equivalent to Lockheed's Skunk Works). The executive confirmed what I had earlier learned from an intelligence insider: that Boeing had teamed up with Europe's Airbus Industrie to manufacture the Nautilus.*

8) *The TR3-A 'Pumpkinseed' is a super-fast air vehicle. The 'Pumpkinseed' nickname is a reference to its thin oval airframe whose contours resemble that seed. It is the craft identified as using Pulse-Detonation Wave Engine (PDWE) technology for propulsion in a sub-hypersonic regime. The Pumpkinseed then uses antigravity technology for reducing the vehicle's mass and for field propulsion at higher altitude and speed levels. As air breathers, these Pulse Detonation Wave Engines can propel a hypersonic aircraft towards Mach 10 at an altitude in excess of 180,000 feet. When used to power an trans-atmospheric vehicle, the same PDWEs are capable of lifting the craft to the edge of space when switched to rocket mode.*

9) *The TR3-B 'Astra' is a large triangular anti-gravity craft within the secret U.S. spacefaring fleet. Black-projects defense industry insider Edgar Rothschild Fouche' wrote about the existence of the TR3-B in his book Alien Rapture. The TR3-B does not depend solely or principally on its hydrogen-oxygen rockets. It is a highly-reduced-gravity aerospace craft manufactured in secret "black programs" by Boeing. The reduced-gravity field it produces reduces the vehicle's weight by about 90% so that very little thrust is required to either keep it aloft or to propel it at speeds of Mach 9 or higher.*

*The TR-3B vehicle's outer coating is electrochemical-reactive and changes with electrical radio-frequency radar stimulation, and can change reflectiveness, radar absorptiveness, and color. This is also the first US vehicle to use quasi-crystals in the vehicle's skin. This polymer skin, when used in conjunction with the TR-3B's Electronic Counter Measures and Electronic Counter-Countermeasures (ECCM), can make the vehicle look like a small aircraft, or a flying cylinder - or even trick radar receivers into falsely detecting a variety of aircraft, no aircraft, or several aircraft at various locations!*

*A circular plasma-filled accelerator ring called the Magnetic Field Disrupter [MFD] surrounds the rotable crew compartment and is far ahead of any imaginable technology. Sandia and Livermore National laboratories developed the reverse-engineered MFD technology. The mercury-based plasma is pressurized at 250,000 atmospheres at a temperature of 150 degrees Kelvin, and accelerated to 50,000 rpm to create a super-conductive charged plasma with resulting gravity-disruption [reduction of almost all of the pull of gravity and effects of inertia].*

*The MFD generates a magnetic-vortex field which disrupts or neutralizes the effects of gravity by 89 percent on a mass within proximity. The MFD creates a disruption of the Earth's gravitational field upon the mass within the circular accelerator. The mass of the circular accelerator and all mass within the accelerator, such as the crew capsule, avionics, MFD systems, fuels, crew environmental systems, and the nuclear reactor, are reduced by 89%. The current MFD in the TR-3B craft causes the effect of making the vehicle extremely light, and able to outperform and outmaneuver any craft yet constructed - except of course those back-engineered total-antigravity craft, which the government does not admit exist.*

*The TR-3B is a high-altitude, stealth reconnaissance platform with an indefinite loiter time. Once it gets to operational altitude, it doesn't take much propulsion to maintain altitude. With the vehicle mass reduced by 89%, the craft can travel at Mach 9 vertically or horizontally. My sources say the performance is limited only the stresses that the human pilots can endure. Such stresses are greatly reduced, considering that along with the 89% reduction in mass, the inertial G (gravity) forces are also reduced by 89%. The crew of the TR-3B can comfortably take up to 40Gs.*

*The TR-3Bs propulsion is provided by three multimode thrusters mounted at each bottom corner of the triangular platform. The TR-3 is a sub-Mach 9 vehicle until it reaches altitudes above l20,000 feet - then above atmospheric drag it can go much faster!*

*The reactor heats the liquid hydrogen and injects liquid oxygen into the supersonic nozzle, so that the hydrogen burns concurrently in the liquid- oxygen afterburner. The multimode propulsion system can operate in the atmosphere with lift from the Magnetic Field Disrupter powered by the nuclear reactor and propulsion by burning hydrogen, and in orbit it uses the combined hydrogen/oxygen propulsion. The engines are reportedly built by Rockwell.*

10) The Northrop antigravity disc (designation unknown) is manufactured by Northrop Grumman's Advanced Concepts and Technologies Division. I have dubbed it the 'Great Pumpkin' from its brilliant ruddy golden-orangish glow. I first saw these craft operationally test-flown in 1992 above the Groom Range ridge line at Area 51, Nevada. Later I saw the same intensely burning-bright orange-gold craft that I had seen above Areas 51 being test-flown sixty miles north of Los Angeles in the Tehachapi Mountains northwest of Edwards Air Force Base.

There Northrop has its secret saucer manufacturing works buried deep within Tehachapi Mountain. I saw the same intensely burning-bright orange-gold craft test-flown above Northrop's ridge-top test bed there as I had seen flown above Area 51(11). When energized, these discs emit their characteristic intense glow. It is reasonable to assume that this is due to strong ionization, and that electrogravitics is the methodology of their field propulsion.

11) The XH-75D (or XH Shark) antigravity helicopter is manufactured by Teledyne Ryan Aeronautical Corporation of San Diego (now part of Northrop-Grumman). USAF Colonel Steve Wilson reported that many of these XH-75Ds were assigned to the Delta/National Reconnaissance Organization Division which retrieves downed UFOs. Rogue elements within that Division, controlled by the Cabal, are also implicated in mutilating cattle as a psychological warfare program on the American public to try to get citizens to fear and hate extraterrestrials through assuming that aliens in antigravity craft are the ones cutting up their cattle.

The XH-75D is also obtained and used by the Cabal for MILABS kidnappings of innocent civilians, who are drugged, hypnotized, and flown off in these silent antigravity craft and given the impression that they are aboard a "flying saucer". Colonel Wilson USAF leaked the existence of the XH-75D "Shark".

12) The Northrop Quantum Teleportation Disc. Are the above antigravity-field propulsion craft the current state-of-the-art in advanced aerospace craft? No. There have been advances beyond "mere" antigravity field propulsion. Quantum physics is now being used to update a variety of aerospace craft and their guidance systems.

On a Sept. 16, 2005 field trip to the boundary of Area 51, during a middle-of-the-night observation, I saw first one, then another, and finally six brightly-lit objects suddenly appear at approximately 1000' (305 meters) height above the desert floor. The intensely-glowing, ruddy, golden-orangish ionization field surrounding these craft appeared identical to the field around the Northrop antigravity disc. But in the 13 years since I had last observed the Northrop discs above Area 51, and at their Tehachapi Mountains manufacturing site, considerable progress has been made.

*As of 1992 a Northrop antigravity disc could slowly rose vertically from its flight pad and gradually reached flight altitude. But in 2005, the Northrop Quantum Teleportation Discs are able to instantly wink out (depart) from their flight pad and suddenly appear at flight altitude without any visible ascent. And it is not a matter of their ionization field having been turned off during ascent for stealth purposes. The ionization field accompanies the electrogravitic field propulsion. If the ionization were turned off, the craft would have fallen from the sky. Rather what apparently is going on is that the Northrop engineers have incorporated quantum-physics principles into the propulsion. Simply stated, Northrop appears to have harnessed Quantum Entanglement to achieve Quantum Teleportation. To the observer, the craft appears to simply cease to exist on the flight pad and instantly begin to exist at say 3000 feet altitude. If the interpretation of this observation is correct, then there exists an 12th entry in the U.S. antigravity arsenal: the Northrop Quantum Teleportation Disc.*

*If the black-budget scientists keep advancing along these lines, we could foresee the day when a fleet of Air Force craft suddenly "wink out of view" on an air base runway and instantly appear at 35,000 feet altitude over a target city halfway around the globe, using quantum principles of Non-Locality and Entanglement. Surprise!*

Aliens and Secret Technology

Chapter 17: The Secret Space Force

Claims of a Secret Space Force and interplanetary bases maintained by the countries of Earth is one of the latest subjects about top secret efforts. An interesting proposition, but evidence is hard to come by.

Here is one story, and so far the only detailed story I can find details of a secret space force:

> NASA hacker: I found evidence America has Deep Space Warships
>
> According to a hacker who is facing a ten-year legal battle after breaking into NASA computers, the United States has a fully operational fleet of Space Warships. Gary McKinnon, firmly believes that he came across the ultimate information that proves the US have a secret space program operated by the navy with fully operational warships that operate in Space.
>
> In a new interview on UFO channel Richplanet TV, McKinnon finally reveals the entire truth about his findings saying: 'I kept going for months and months. I kept thinking, 'They're going to close this door'.
>
> McKinnon said that he used a software called Landsearch which allowed him to search all files and folders of interest to him. 'I scanned and looked for documents, I found an Excel spreadsheet which said, 'Non-terrestrial officers', states McKinnon. 'It had

ranks and names. It had tabs for 'material transfer' between ships. 'I took that to be, they must have a ships based in space – the names started with U.S.S.'

Gary McKinnon is accused of mounting the biggest ever hack in the history of the United States by breaking into the computers of the Army, Air force, Navy and NASA.

During his latest interview, McKinnon describes a conversation from a former NASA whistleblower Donna Hare, who had been told by a colleague that NASA was trying to hide delicate information by "airbrushing" UFOs from their photos.

"There was a colleague, who was in another room — they all had secret clearance but they were on different projects — and she (Hare) was in this chap's lab or room or whatever it is and he said come and take a look at this," said McKinnon.

While there are many people who firmly believe this is yet another elaborate hoax, there have been a series of former NASA and government employees who back up the claims made by McKinnon.

According to statements from a former Marine not only have humans made it to Mars in the past, but we have developed a secret space program and flotilla that operates in space. The former US Marine, was posted on the red planet for years and his mission was to protect the five human colonies from indigenous life forms on Mars. According to the former Marine known as Captain Kaye, not only did he spend years on Mars, but he also served aboard a giant space carrier for three years. He worked for the Mars Defense Force (MDF) which is owned and operated by the Mars Colony Corporation (MCC) which is basically a conglomerate of financial institutions, government and tech companies. Kaye and his team were part of a special section of the United States Marines with a highly classified mission, protect and ensure the existence of five newly-established colonies on the surface of the red planet. The Earth Defense Force, another

> secret military branch has military recruits from countries such as the United States, China and Russia.
>
> Parts of the testimony from Captain Kaye are consistent with that of Michael Relfe, another whistleblower who claims to have served 20 years tour on the red planet. Laura Magdalene Eisenhower, great-granddaughter of former President Eisenhower claims that efforts were made to recruit here onto a human colony on Mars, led by researcher Dr. Hal Puthoff.
>
> According to "Jackie", there are humans on Mars, and we have been on the red planet for over 20 years. 'Jackie', she and six other employees saw the exact same thing, suggesting this might be the evidence that proves that a secret space program did exist (or might even still).
>
> According to 'Jackie' while working as part of the team downloading telemetry from the Viking Lander, she saw human setting foot on the surface of the red planet via a live feed from Mars.
>
> In an interview with Coast-to-Coast AM, Jackie talked about humans on Mars, something that has been discussed among other alleged NASA employees for years.
>
> Many Ufologists firmly believe that the statements made by the above individuals is accurate and that this information is being withheld from the general public.

Numerous claims have been made in recent years that support the result of one of the biggest security breaches In the history of the US military. According to reports, the US Navy and NASA have a fully operational Space Fleet. Interestingly, these 'accusations' are backed up by highly ranking individuals.

Finally, having worked on classified products for a number of years I can tell you that secret and top secret and above information all reside on separate networks with that information only being viewable in those closed areas. It does seem odd to me that something as highly classified as a secret space force would have names about the crews of those ships sitting on an unclassified computer which could be hacked from the internet.

There are some stories that we have established a secret moon base as far back as the early 1960s. A project was supposedly approved to build a U.S. secret moon base. This excerpt is from the book "The Secret Space Program and Other Space Coverups" (17):

> On March 20, 1959, Lieutenant General Arthur G. Trudeau, then U.S. Army Chief of Research and Development, signed off on Project Horizon, the establishment of a U.S. Army outpost on the Moon, at a cost of approximately $6 billion. It would be a permanently manned base that would demonstrate decisive military superiority over the former Soviet Union. The Army's goal was to have the initial stages of the program in place by 1965-the very same year that Wolfe was told about NASA evidence of an installation.
>
> In the opening pages of the several hundred-page report entitled Project Horizon: A U.S. Army Study for the Establishment of a Lunar Military Outpost, Trudeau wrote:
>
> "There is a requirement for a manned military outpost on the moon. The lunar outpost is required to develop and protect potential United States interests on the moon; to
>
> develop techniques in moon-based surveillance of the earth and space, in communications relay, and in operations on the surface of the moon; to serve as a base for exploration of the moon, for further exploration into space and for military operations on the moon if required; and to support scientific investigations on the moon."
>
> It is true that Project Horizon went "black" and the military continued a super-secret space program about which NASA

*knows very little or nothing at all? Hence the space agency's deep concerns relative to the mysterious facility photographed on the far side of the Moon back in the 1960s.*

Aliens and Secret Technology

Aliens and Secret Technology

Chapter 18: The United States incredible technology

To give you an idea of how much advanced the United States is in aerospace technology above what the world thinks, some quotes make interesting reading:

Ben Rich the second director of the Lockheed Skunkworks which built many classified planes including the SR-71, the U-2, and F117 Stealth fighter. Here are some comments he made:

> "Ben Rich told me twice before he died: 'We have things at Area 51 that you and the best minds in the world won't even be able to conceive that we have for 30 or 40 years, and won't be made public for another 50." A friend of mine at Lockheed told me: "We have things in the Nevada desert that are alien to your way of thinking far beyond anything you see on Star Trek."

Another Master Sergeant Interview:

> A master sergeant was interviewed who had been at Groom Lake three different times as an Air Force safety specialist. ... At first he was real nervous, but when he warmed up he told me: 'We have things that would make George Lucas envious.' I know one retired guy who worked at Lockheed for 30 years, most of the time at Area 51; he's very proud of what he's done, and he wants the story of the place to be told so that his grandchildren will have some idea of what he was involved in. In the summer of '86 I asked him if he believes in UFOs. He said, 'They absolutely, positively do exist!' I said, 'Can you expand on that?' And he said, 'No, I've said too much as it is.' "

Another quote by Ben Rich:

> As Ben Rich said, who developed the stealth aircraft back in 1993 in the presence of two friends of mine: "We now have the technology to go to the stars". And that was back in 1993.' When Mr. Good was asked to elaborate on where this technology came from, he replied: 'Alien crafts [sic]. The study of alien crafts (craft)

> *that have been recovered and by liaising with actual aliens who have helped us develop this technology.*

The point of these quotes is that the United States government is keeping aerospace and space technology top secret which would give us a huge military advantage either in the atmosphere or in orbit , and even out in the Solar System.

Given that the United States public military budget already equals that of the rest of the worlds combined, the U.S. has an incredible technical and qualitative edge in technologies for war.

Here is a picture of Ben Rich:

The United States already dominates the plant Earth. So what else would we use this super-secret technology to do?

It would explain the development of a Secret Space Force using the ships and technologies we have already developed to defend Earth from Aliens.

Aliens and Secret Technology

Chapter 19: Summary-My Theory of aliens and secret technology

Lots of the information presented in this book can be argued as valid or invalid. I tend to use an old intelligence technique which is that one can put together classified information from using enough unclassified sources to generate a real picture of the secret reality.

My conclusions about Aliens, UFOs, and secret technology are as follows:

1) UFOs and aliens have been seen throughout human history. There is evidence of sightings and interactions with aliens from the dawn of history and even before man existed. Aliens and UFOs are real and we need to learn to accept them and work with them.
2) Some type of anti-gravity technology became available to the United States soon after World War Two. Whether it was a result of Nazi research on the Bell, the alien ship found at Roswell, or Townsend-Brown anti-gravitics—I really don't know the truth. But anti-gravity technology clearly exists and has been the subject of U.S. secret research since then.
3) The weapons in the arsenal of the United States are almost beyond comprehension. We would not survive a world-wide nuclear war, but nobody else would either. No other country has a chance to attack the U.S. and survive.
4) Some top secret aircraft definitely exist—The Pumpkin Seed, and TR3B as examples and are part of our secret aerospace fleet.
5) Our governments and militaries have likely had a working relationship with aliens since at least the late 1940s. Some trips by humans in their spacecraft have probably resulted.
6) Aliens are living on Earth—maybe for thousands of years. Some live in plain site because they look exactly like us. Some can use makeup to blend in, and some use invisibility technologies to avoid being seen.
7) Our Earth based governments have probably identified over one hundred alien races from various interactions here on our planet.
8) Anti-Gravity and Zero Point technologies are being withheld from the public for a couple of reasons. One main reason is potential

disruption of large industries like the worldwide oil and gas industry. Another reason is that government authorities don't want to panic the general population about the existence of aliens who visit our planet too.

9) Assuming we have had anti-gravity ships since the early 1960s, then it would be a natural growth path to also assume we have military bases on the moon and possibly Mars.

So what should we conclude about all of the claims and counter claims about aliens, UFOs, secret anti-gravity, and secrete zero point technologies? This field is full of conspiracy theories and lots of contradictory information after all. There is probably an incredible amount of disinformation we are dealing with too.

I should also mention that I went to the 2017 "Contact In the Desert" UFO conference out in the California desert near Joshua Tree National Park. I attended a number of presentations looking for facts and evidence. Unfortunately, I have to say that many of the persons presenting either were mentally ill, or total scam artists taking advantage of the gullible.

However, there were a few presenters like Steven Greer who were sincere researchers of UFOS and aliens who had many interesting facts to share.

There are now enough credible witnesses of UFOs, and/or secret government technologies that we can be sure something extraordinary really exists.

It's also likely that with the number of habitable planets we are learning about aliens are very common in our galaxy and many have been to Earth and made contact with our governments.

The good news about all of this is that we have an exciting future to look forward to. We will be able to settle the Solar System and travel to the stars. We will also soon have interstellar commerce and relationships with many alien races.

The future is going to be amazing !

Aliens and Secret Technology

## Bibliography

1. Drake Equation. *Wikipedia.* [Online] [Cited: 5 30, 2009.] http://en.wikipedia.org/wiki/Drake_equation.

2. Age of the Universe. *Wikipedia.* [Online] [Cited: 5 30, 2009.] http://en.wikipedia.org/wiki/Age_of_the_universe.

3. Age of the Earth. *Wikipedia.* [Online] [Cited: 5 30, 2009.] http://en.wikipedia.org/wiki/Age_of_the_Earth.

4. 'Invisibility Cloak' Successfully Hides Objects Placed Under It. *Science Daily.* [Online] [Cited: 5 30, 2009.] http://www.sciencedaily.com/releases/2009/05/090501154143.htm.

5. https://www.express.co.uk/news/weird/761676/ALIEN-picture-Parque-Forest-Santiago-Chile-Carabineros. *www.express.co.uk.* [Online] 02 07, 2017.

6. Open Source Intelligence. *Wikipedia.* [Online] [Cited: 5 31, 2009.] http://en.wikipedia.org/wiki/Open_Source_Intelligence.

7. African Gods According to the Dogon. *Church of Critical Thinking.* [Online] [Cited: 6 4, 2009.] http://churchofcriticalthinking.org/dogon.html.

8. http://www.dailymail.co.uk/news/article-4252164/Files-suggest-Nazis-tested-NUCLEAR-BOMB-WW2-ended.html. *The Daily Mail.* [Online] 02 03, 2017.

9. Farrell, Joseph P. *SS Brotherhood of the Bell: The Nazis' Incredible Secret Technology.*

10. http://www.ufos-aliens.co.uk/cosmicunder.html. *ufos-aliens.co.uk.* [Online]

11. Cook, Nick. *The Hunt for Zero Point Energy.*

12. Serpo.Org. *http://serpo.org.* [Online]

13. http://www.bibliotecapleyades.net/ciencia/ciencia_flyingobjects11.htm. *bibliotecapleyades.net.* [Online]

14. Hall, Charles James. Millennial Hospitality. s.l. : 1st Books, 2002.

15. Greer, Steven M. *Disclosure.* s.l. : Carden Jennings Publishing, 2001.

16. www.ufos-aliens.co.uk. http://www.ufos-aliens.co.uk/cosmicspecies.htm. [Online]

17. Patton, by Olav Phillips & Ron. *The Secret Space Program and Other Space Coverups.* s.l. : Paranoia Publishing.

18. Authors, Many. *The Holy Bible--NIV, KJV, NIS, LB Versions.* Various.

19. Various. *The Holy Bible.* s.l. : KJV, NIS, IS, RSV.

# Index

100 million planets, 14
age of our Universe, 14
Alien Races, 69
Alien related paranormal experiences, 23
Aliens on the Astral Plane, 24
Ancient grooved spheres, 27
Anunnaki, 28
Astra, 78
Aurora, 75
B-2 Stealth Bomber, 74
Ben Rich, 89
Biefeld–Brown effect, 45
CARET Programming Diagram, 63
Crop Circles, 23
Deep Space Warships, 83
Disclosure, 67
Dogon Tribe, 33
doughnuts on a rope, 72
Dr. Richard Boylan, 74
Draconians, 69
Dulce, New Mexico, 53
Edgar Fouche, 73
Electrogravitics, 45
Ezekiel and the flying machine, 30
F-22 Raptor, 74
F-35 Lightning II, 75
Foretelling the Future, 21
fresco entitled "The Crucifixion, 32
Full Drake Equation, 13
Ground Based Missile Defense, 11
Hans Kammler, 43
invisible alien, 16
Legends of the Annunaki, 28
Lockheed Skunkworks, 89
Lockheed X-22A, 76
Lockheed-Martin X-33A, 76
Nick Cook, 55
Nommos, 33
Northrop antigravity disc, 80
Northrop Quantum Teleportation Disc, 81
Operation Paperchip, 42
OSINT, 18
Out of Body Experience, 21
Paintings on Stone walls, 30
possible underground tunnels, 54
Psychic Healing, 21
Pumpkin Seed Hypersonic Craft, 71
Pumpkinseed, 78
Puritans in New England, 35
Robert Scott Lazar, 73
Roswell Incident, 47
S-4, 73
Scale of Believability, 19
Sergeant Robert Dean, 65
Sightings in the Nineteenth Century, 38
Steven M. Greer, 67
The Day After Roswell, 49
The Drake equation, 13
The Greys, 69
The Hunt for Zero Point, 55
The Nautilus, 77
The Nazi Bell, 41
The Nordics, 69
The Palo Alto Caret Laboratory, 61
The Secret Space Force, 83
The Serpo project, 57
Thomas Townsend Brown, 45
TR3-A, 78
TR-3B, 72
XH-75D, 81
Zero Point Energy, 55
Zeta Reticuli, 60

www.ingramcontent.com/pod-product-compliance
Lightning Source LLC
Chambersburg PA
CBHW040318220526
45473CB00009B/2477